R. Matheson

Special report on Insects, Fungi and Weeds injurious to Farm

Crops in Ireland

R. Matheson

Special report on Insects, Fungi and Weeds injurious to Farm Crops in Ireland

ISBN/EAN: 9783742801333

Manufactured in Europe, USA, Canada, Australia, Japa

Cover: Foto ©Klaus-Uwe Gerhardt /pixelio.de

Manufactured and distributed by brebook publishing software
(www.brebook.com)

R. Matheson

Special report on Insects, Fungi and Weeds injurious to Farm

Crops in Ireland

SUPPLEMENT TO THE AGRICULTURAL STATISTICS
OF IRELAND FOR THE YEAR 1889.

SPECIAL REPORT

ON

INSECTS, FUNGI, AND WEEDS

INJURIOUS TO FARM CROPS.

ILLUSTRATED WITH ORIGINAL DRAWINGS

BY

ROBERT E. MATHESON,

BARRISTER-AT-LAW,

ASSISTANT REGISTRAR-GENERAL AND SECRETARY OF THE GENERAL REGISTER OFFICE.

Presented to both Houses of Parliament by Command of Her Majesty.

DUBLIN:
PRINTED FOR HER MAJESTY'S STATIONERY OFFICE,
BY
ALEXANDER THOM & CO. (Limited),
And to be purchased, either directly or through any Bookseller, from
Eyre and Spottiswoode, East Harding-street, Fetter-lane, E.C., or 32, Abingdon-street,
Westminster, S.W.; or Adam and Charles Black, 6, North Bridge, Edinburgh; or
Hodges, Figgis, and Co., 104, Grafton-street, Dublin.

1890.

TO

HIS EXCELLENCY LAWRENCE, EARL OF ZETLAND,

&C., &C., &C.,

Lord Lieutenant-General and General Governor of Ireland.

MAY IT PLEASE YOUR EXCELLENCY,

Referring to my Annual Report on the Agricultural Statistics of Ireland, bearing date 19th May, I have now the honour to lay before your Excellency the "Special Report on Insects, Fungi, and Weeds Injurious to Farm Crops" referred to therein, which has been prepared by Mr. Robert E. Matheson, B.L., Secretary to this Department, and kindly placed by him at my disposal for public use.

Mr. Matheson has been most courteously assisted by various eminent scientific friends, who are named in his letter, to whom my best acknowledgments are also due.

> I have the honour to be
> Your Excellency's faithful Servant,
>
> (Signed),   THOMAS W. GRIMSHAW,
> Registrar-General.

General Register Office,
Charlemont House,
Dublin, September, 1890.

[v]

To Thomas W. Grimshaw, Esq., M.D.,
      Registrar-General.

Dear Sir,

I have pleasure in placing at your disposal, for use in connection with the Agricultural Statistics, a Treatise by me on Insects, Microscopic Fungi, and Weeds Injurious to Farm Crops.

The idea of such a paper was first suggested to my mind on inspecting the interesting collection of insects made by Mr. S. E. Mosley, of Huddersfield, and purchased from him by the authorities of the Science and Art Museum in Dublin. It occurred to me that if I could present in a short paper the main facts concerning the insects noxious to crops in this country I would perform a useful service to the Irish agriculturist. As the project became more matured, I decided to include also notices of the principal microscopic fungi, and of the weeds which cause trouble to the farmer.

As regards the insects, I have taken as the basis of my remarks the very valuable and succinct descriptions supplied by the authorities of the Museum ; also availing myself of the information given in the works of the eminent authorities, Mr. Curtis, Miss Ormerod, and Mr. Whitehead.

For the Fungi, I have referred to and freely used the works of two eminent mycologists, Dr. M. C. Cooke and Mr. Worthington Smith, and for the Weeds I have similarly used Bentham's British Flora.

I have to acknowledge the courtesy of Dr. V. Ball, Director of the Science and Art Museum, and of Dr. Scharff, Curator of the Natural History Department, in kindly affording me special facilities for my work.

My best thanks are due to my friend, Mr. G. H. Carpenter, Entomologist in the Science and Art Museum, for practical help in preparing the portion of the paper relating to Insects, and assisting me with valuable information as to the records of the insects found in Ireland.

As regards the botanical portions, I am indebted to the kindness and courtesy of Dr. E. Perceval Wright, Professor of Botany in the University of Dublin, for revising the chapter on Fungi, and furnishing me with lists of native plants ; also to Mr. Greenwood Pim, M.A., for very kindly revising the chapter on Fungi ; and to Mr. Richard M. Barrington, LL.B., for supplying me with valuable information regarding those weeds troublesome to agriculturists in Ireland.

I desire also to offer you my best thanks for your own assistance and advice as regards the chapter on Weeds, which your experience as former Professor of Botany in the Queen's University enabled you to give me.

The illustrations are from original drawings made by me of the specimens in the Museum, the figures of the smaller insects being drawn from the microscope. In this portion of the work I have been materially aided by Mr. Carpenter and also by my wife.

The paper has been designedly prepared on the simplest lines, and the use of scientific terms avoided as far as possible.

Trusting it may prove acceptable and helpful to persons interested in agriculture in Ireland,

I am, dear Sir,

Yours very faithfully,

(Signed),   ROBERT E. MATHESON.

Primrose Hill,
    Kingstown, May, 1890.

TABLE OF CONTENTS.

INTRODUCTORY REMARKS.

THE object of this paper is to bring before Irish agriculturists the facts ascertained by Scientists regarding their three enemies — Insects, microscopic Fungi, and Weeds—to give them the material for detecting these foes, and to place before them the best methods of dealing with them.

Nature has linked together these three branches in a remarkable way. There is a direct connection between Insects and Weeds, as certain insects frequent plants of particular botanical families, and the weeds of these families, if left undestroyed, shelter them during the winter till, in the spring, they go forth afresh on their work of destruction. Again, Fungi stand in intimate relation to weeds, as certain kinds of Fungus are found on plants of particular botanical orders, and the weeds belonging to these families nurse the spores of the Fungus, which sow themselves when winter is over on the adjacent crops. Further, Insects occupy an interesting position with regard to Fungi, as it has been conclusively shown that these creatures carry the spores of Fungus on their wings and bodies, and thus disseminate the seeds of disease all around.

Some elementary knowledge of the various Orders of Insects and their life-histories is a necessary introduction to the prevention of their ravages to crops. For it frequently happens that injurious insects are killed and devoured by other insects. Ignorant persons destroy the latter as well as the former, and thus make war on their own friends and allies. Again, a knowledge of the stages through which insects pass in the course of their development or metamorphosis is of great importance, as the ravages committed vary in the different stages.

All insects are hatched from eggs[*]. Some, like the grasshoppers, leave the egg in a form very like that of their parents, except that they have no wings. In this state the young insect is called a *Larva*. It grows and undergoes successive changes of its skin, until it becomes a *Nymph*, when the first beginnings of wings appear. After a final casting of the nymph-skin, the perfect winged grasshopper is produced. All through these changes, which make up what is called an *incomplete metamorphosis*, the creature moves and feeds.

But in the case of other insects, as the Butterflies, the young when hatched do not seem to resemble their parents at all.

[*] Occasionally the eggs are hatched within the body of the mother, and the young born alive.

They begin life as *Larvæ* or caterpillars, which grow very fast, and often change their skins in the course of growth. When fully grown, the caterpillar changes to a *Pupa* or chrysalis, which remains almost or quite motionless for a considerable time, at the end of which its skin splits and the perfect winged butterfly comes out. This kind of growth is called *complete metamorphosis.*

Now it is of great importance to remember that only in the perfect final condition does an insect lay eggs. People ignorant of this elementary fact have been known to destroy the cocoons of Ichneumon flies, thinking them to be "caterpillars' eggs," instead of what they really are, the coverings of the caterpillars' greatest enemies.

Naturalists divide the true insects into two great groups, according to the nature of their metamorphosis. Each of these groups is again divided into several orders. The leading facts about these orders, with special reference to their bearing on insect ravages, is here summarized.

A.—*Insects with incomplete metamorphosis.*

1. *Order Orthoptera.*—Under this order are included those insects with an incomplete metamorphosis, which have their jaws formed for biting, and their front pair of wings developed as a hard covering for the hind pair, which are gauzy. The order comprises the Earwigs, Cockroaches, Grasshoppers, Locusts, and other insects, most of which destroy vegetation in all the stages of their growth.

2. *Order Pseudoneuroptera.*—Under this head are placed insects like the Dragon Flies and May Flies, which resemble the Orthoptera in their biting jaws, but which have all four wings gauzy and generally transparent. The insects of this order are usually to be regarded as allies, for instead of devouring vegetation, most of them, in all stages of growth, hunt and feed on smaller insects.

3. *Order Rhynchota.*—These insects vary much in the nature of their wings, but they all agree in the structure of the mouth, which is formed for piercing and sucking, not as in the two preceding orders for biting. Some of the most destructive of insects belong to this group, as the Plant Bugs, Plant Lice, Scale Insects, and others. The Plant Lice (*Aphidæ*) attack all kinds of vegetation, and their very small size and immense numbers make it difficult to check their ravages.

B.—*Insects with complete metamorphosis.*

4. *Order Neuroptera.*—This order contains those insects with complete metamorphosis which have two pairs of gauzy wings

and biting jaws. Here are the Caddis Flies and others which are generally not of much importance to crops either as friends or foes, for the larvæ often live under water. The Golden Eyes, however, whose grubs feed on Plant Lice, belong to this order.

5. *Order Coleoptera.*—This very important order includes the Beetles—insects with biting jaws and with the front pair of wings transformed into hard covers for the hind pair, which alone are used for flight. The eggs laid by beetles hatch out grubs, some of which feed in wood, others underground on roots, others on leaves, others on small insects, and others on carrion. The grub when full grown turns to a pupa, which lies quietly in wood or underground until the perfect beetle is produced. The habits of beetles are as various as those of their grubs.

Specially destructive families are the Weevils (which may be easily recognized by the snout-like prolongation of the head), the Click Beetles or Skipjacks (whose grubs known as "wire worms" devour roots), and the Leaf Beetles (small generally metallic species feeding on leaves).

On the other hand, the Lady-Birds are valuable allies, and should be preserved, as they and their grubs feed on Plant Lice. The Carnivorous Ground Beetles and Tiger Beetles (to be known by their prominent and powerful mandibles and longish slender legs) are also generally to be encouraged.

6. *Order Diptera.*—This order includes the two-winged Flies, all of which have the hinder pair of wings reduced to little knobs. Their jaws are formed for piercing. They begin life as grubs or maggots which often cause great injury to vegetation. The insect pests which attack cattle, horses, and sheep mostly find their place here. Some *Diptera*, however, are beneficial. The grubs of the Hovering Flies (*Syrphidæ*)—flies with yellow-banded, wasp-like bodies—devour the Plant Lice ; whilst the grubs of some Flies related to the House-fly feed within the bodies of Caterpillars.

7. *Order Lepidoptera.*—The Butterflies and Moths which make up this order have their four wings covered with scales, producing beautiful patterns. The mouth is formed only for sucking. The caterpillars from which they develop have jaws for biting and generally feed on leaves, sometimes in wood. The Caterpillars of the White Cabbage Butterflies and of many species of Moths attack crops of various kinds. Many caterpillars, however, feed on useless or noxious weeds.

8. *Order Hymenoptera.*—The insects of this order have generally four transparent wings, and their jaws are adapted both for biting and licking.

They all develop from maggots or grubs, but their manner of

feeding varies greatly in different families. The Saw Flies, which pierce leaves and wood in order to lay eggs, are very destructive, the grubs often causing great havoc when hatched. The Gall Flies, which cause a peculiar growth or "gall" on the plants within which the grubs feed, are also injurious, though not so bad as the Saw Flies. The Ichneumon Flies, on the other hand, which lay their eggs within the bodies of caterpillars and other insects where the grubs feed, causing the death of the creature attacked, are among the farmers most useful allies.

Other well-known insects belonging to the *Hymenoptera* are Ants, Wasps, and Bees.

Though not strictly speaking Insects, Millipedes and Mites are referred to in this paper. The Millipedes belong to the *Myriapods* or " many legs," which are divided into two orders, the *Millipedes* and the *Centipedes*. The Millipedes, which have rounded bodies and two pairs of legs on each segment, are vegetable feeders and injurious. The *Centipedes*, however, which have flattened bodies and only one pair of legs on each segment, are carnivorous and altogether beneficial. The Mites are (usually) little creatures with four pairs of legs, and the whole body looks as if made in one piece. This distinguishes them from the Spiders (with which they are often confounded), which have the hinder-body sharply marked off from the forward part. Many Mites are injurious, but all true Spiders are insect eaters and beneficial.

The second part of this treatise deals with the microscopic Fungi which cause rust, smut, mildew, and mould in crops of various kinds.

They are minute vegetable organisms, parasitic in their character, which reproduce themselves in immense numbers by means of spores.

It would be beyond the scope of this paper to enumerate all the Fungi which are known to attack agricultural produce, and only, therefore, those forms most frequently met with will be considered.

The Fungi have been arranged under the different classes of crops upon which they are found, and the nature of the remedies available described.

Fig. 1.
ANTLER MOTH.
(*Charaeas graminis.*)

Fig. 2.
CLOUDED BORDER BRINDLE MOTH.
(*Xylophasia rurea.*)

Fig. 3.
GHOST SWIFT MOTH.
(*Hepialus humuli.*)
[Male.]

Fig. 4.
RUSTIC SHOULDER KNOT MOTH.
(*Apamea basilinea.*)

Fig. 5.
DARK ARCHES MOTH.
(*Xylophasia monoglypha.*)

Fig. 3 (a)
[Female.]

Fig. 6.
LARGE YELLOW UNDERWING MOTH.
(*Tryphaena pronuba.*)

Fig. 5 (a).
CATERPILLAR.

Fig. 4 (b).
CATERPILLAR.

Fig. 3 (a).
CATERPILLAR.

Fig. 6.
CORN WOLF MOTH.
(*Charaeas graminis.*)

Fig. 5.
COMMON SWIFT MOTH.
(*Hepialus lupulinus.*)

FIG. 1.
SNOWY FLY.
(*Aleyrodes proletella.*)

(Mag.)

(Nat. size.)

FIG. 6.
FLOWER BEETLE.
(*Meligethes aeneus.*)

(Nat. size.)

(Mag.)

FIG. 2.
SEED WEEVIL BEETLE.
(*Ceutorhynchus assimilis.*)

(Nat. size.)

(Mag.)

FIG. 3.
ROOT GALL WEEVIL BEETLE.
(*Ceutorhynchus sulcicollis.*)

(Nat. size.)

(Mag.)

FIG. 5.
MUSTARD BEETLE.
(*Phaedon cochleariae.*)

(Nat. size.)

(Mag.)

FIG. 4.
FLEA BEETLE.
(*Phyllotreta nemorum.*)

(Nat. size.)

(Mag.)

FIG. 7.
ROOT FLY.
(*Anthomyia radicum.*)

FIG. 8.
BLACK LEAF MINER FLY.
(*Phytomyza nigricornis.*)

(Mag.)

(Nat. size.)

FIG. 9.
YELLOW LEAF MINER FLY.
(*Drosophila flava.*)

FIG. 10.
TURNIP SAW FLY.
(*Athalia spinarum.*)

FIG. 11.
CATERPILLAR.

PLATE 19.

Fig. 1.
LARGE WHITE BUTTERFLY.
[Pieris brassicae]
[Male.]

Fig. 1 (a).
LARGE WHITE BUTTERFLY.
[Pieris brassicae]
[Female.]

Fig. 2.
SMALL WHITE BUTTERFLY.
[Pieris rapae]
[Male.]

Fig. 2 (b).
CATERPILLAR.

Fig. 2 (a).
SMALL WHITE BUTTERFLY.
[Pieris rapae]
[Female.]

Fig. 3.
RED BAND MOTH BUTTERFLY.
[Pieris rapae]
[Female.]

Fig. 4.
CABBAGE MOTH.
[Mamestra brassicae]

Fig. 5.
COMMON DART MOTH.
[Agrotis segetum]
[Male.]

Fig. 4.
HEART AND DART MOTH.
[Agrotis exclamationis]

Fig. 4 (a).
CATERPILLAR.

Fig. 5 (a).
CATERPILLAR.

Fig. 5.
GARDEN PEBBLE MOTH.
[Pionea forficalis]

Fig. 6.
DIAMOND-BACK TURNIP MOTH.
[Plutella maculipennis]

Fig. 1.
CARROT FLY.
(*Psila rosæ.*)

(Nat. size.)

(Mag.)

Fig. 2.
COMMON FLAT BODY MOTH.
(*Depressaria applana.*)

(Nat. size.)

Fig. 3.
SOLITARY WASP.
(*Odynerus.*)

Fig. 4.
CELERY FLY.
(*Tephritis onopordinis.*)

(Mag.)

(Nat. size.)

Fig. 5.
CELERY STEM FLY.
(*Piophila apii.*)

(Nat. size.)

(Mag.)

Fig. 7.
ONION FLY.
(*Anthomyia ceparum.*)

(Nat. size.)

Fig. 6.
LETTUCE FLY.
(*Anthomyia lactucæ.*)

(Mag.)

FIG. 1.
FROG FLY.
(*Lophorys volans*)

(Mag.)

FIG. 2.
COLORADO BEETLE.
(*Doryphora decemlineata*)

(Nat. size.)

FIG. 3.
HAWK MOTH (Death's Head.)
(*Acherontia atropos*)

FIG. 3 (a).
CATERPILLAR.

PART I.

INSECTS INJURIOUS TO FARM CROPS.

Insects Injurious to Corn, Grass, and Clover.

INSECTS INJURIOUS TO CORN AND GRASS.

Millipede.—Julus sabulosus, Leach.

[PLATE I., FIG. 1.]

These Millipedes, which are of a blackish colour, and other related species, devour the roots of cereals and grass. They lodge beneath clods and stones. They are said to be destroyed by watering with solution of nitrate of soda or lime water. Soot spread on the land will also drive them away.

Another species of Millipede—Polydesmus complanatus—of a light brown colour [Plate I., Fig. 2] is similarly destructive.

The Mole Cricket.—Gryllotalpa vulgaris, Latr.

[PLATE I., FIG. 3.]

These Insects, which are of a brownish colour, devour the roots of plants and live in underground burrows. Ashes, soaked with paraffin, if strewed on the ground will destroy them.

Corn Thrips.—Lisnothrips cerealium, Hal.

[PLATE I; FIG. 4.]

Corn Thrips, which are of a blackish colour, attack ears of corn and suck the juices from the seeds, appearing from June till harvest. The early sown crops are less liable to injury. All borders and hedge bottoms should be cleaned out, as these insects frequent damp places.

Plant Louse.—Aphis granaria, Kirby.

The colour of these insects is green. They suck the sap from the stems and leaves of the young corn, and subsequently attack the ears. Their ravages are greater on the late crops than on those sown early, which have had time to harden before the insects become numerous. The natural foes of the Plant Lice should be protected, and all waste grasses cleared away.

An Ichneumon Fly (Ephedrus plagiator) deposits its eggs inside the Aphis, and its grub devours the insect inside, leaving only a skin. This grub, again, is attacked by a Fly (Megaspilus carpenteri) which lays its egg in the body of the grub. Thus Ephedrus is an ally, while Megaspilus is a foe.

Cockchafer.—Melolontha vulgaris, Fab.

[PLATE I., Figs. 3 and 5 (a).]

The grub of the Cockchafer feeds on roots, living underground. The development occupies about three years. The most effectual enemies of the grubs are the insectivorous birds, such as Starlings, Rooks, and Gulls. The Beetles devour the leaves, and may be destroyed at midday when they settle in the trees by shaking them down. Both grubs and beetles are sought after by pigs.

Other Chafers, such as Phyllopertha horticola (Plate I., Fig. 6), and Rhizotrogus solstitialis (Plate I., Fig. 7), are known to attack crops in a similar manner. There is no record however of the occurrence of the latter insect in Ireland.

Wire-Worm or Click Beetle [Skipjack].—Agriotes lineatus, L.

[PLATE I., Figs. 8 and 8 (a).]

The colour of this Beetle is dull brown. Serious injury is caused to the roots of all kinds of crops by the Wire-Worms or grubs of the Click Beetles. The process of growth occupies some years. When fully grown they go deep down into the ground and change to pupæ, from which Beetles are developed which show themselves in a few weeks above ground. The soil should be carefully examined, and any waste cleared before planting or sowing. Good manuring and heavy rolling are beneficial. Starlings, Plovers, and Rooks, as well as other birds, devour Wire-Worms. After a bad attack the ground should be left fallow for the winter, and then planted with mustard, which the Beetles will not eat. There is another Click Beetle (Agriotes obscurus) resembling the above, but of a darker colour, whose habits are similar.

Rice Weevil Beetle.—Calandra oryzæ, L.

[PLATE I., Fig. 9.]

The colour of this Beetle is dark brown. Corn in granaries is attacked by these insects, which lay their eggs in the grains, and the grubs when hatched eat the seed. The Beetles also devour the grain. The floors of granaries should be scrubbed with soft soap and the walls whitewashed. As frost affects the Beetles, it is advisable when the granary is empty to expose it to the action of the atmosphere in cold weather. Another Weevil Beetle (Calandra granaria), which resembles the foregoing, but is slightly longer, attacks corn in a similar manner.

Ribbon-foot Fly.—Chlorops tæniopus, Meig.

[PLATE I., Fig. 10.]

This Fly is of a yellow or greenish colour, with distinct dark markings. Its eggs are laid at the base of ears of barley and other corn, and the maggots, when hatched, eat the stem. The injury is known as "gout," and may be recognized by the swollen and distorted condition of the stem. Attention to drainage is requisite, the crops being most liable to injury on wet soils. The plants attacked should be taken away and burnt.

Midge.—Cecidomyia tritici, Kirby.

The form of this Insect is well known. It lays its eggs on the flowers of wheat, and the larva sucks the juices of the grain. This is commonly known as the "Red Maggot." Where there has been an attack the chaff should be burnt, or put at the bottom of a dung pit and turned into manure. In this, as in other cases, weeds and wild grasses should be cleared from the vicinity.

Hessian Fly.—Cecidomyia destructor, Say.

When treating of the Midge it may be advisable to insert a notice of the celebrated Hessian Fly (Cecidomyia destructor), which causes immense damage on the Continent and in America, and was first found in England in 1886. This insect appears in April, and lays its eggs on the leaves of the young wheat. The stem is eaten into by the maggots, and thus the ear is deprived of the supply of sap. The stem, being thus weakened, breaks down. The brood of flies appears from the pupæ in August. These lay eggs from which the maggots are produced, which turn to pupæ before the winter. Siftings from crops attacked should be burnt, or a long stubble left and that burnt.

Crane Flies.—Daddy Long Legs—Tipulidæ.

The appearance of these insects is familiar. They lay their eggs on the ground or near the surface in damp grass, in the autumn. The grubs, which appear in spring, are commonly called "Leather Jackets," and eat the underground stems and roots. The pupæ live beneath the ground, and the flies make their appearance in summer. Attention to drainage is necessary, as these flies like damp. The birds which devour insects, such as Starlings, Rooks, and Gulls, are the farmers' friends. The best dressing for crops infested with these insects is said to be guano mixed with salt.

Antler Moth.—Charæas graminis, L.

[PLATE II., FIG. 1.]

The general colour of this Moth is brown; there is a distinct whitish line on the centre of the front wing. The caterpillars frequently occasion much injury to the roots of grasses, more especially in mountain pastures. Rooks and pigs eat them. Watering with lime, or strewing the ground with ashes, has been found beneficial.

Rustic Shoulder Knot Moth.—*Apamea basilinea, F.*

[Plate II., Fig. 2.]

The colour of this Moth is brown. Its eggs are laid in the ears of corn in June. The grains are devoured by the caterpillars which are carried away with the corn. In threshing they may be picked out and destroyed.

Large Yellow Underwing Moth.—*Tryphæna pronuba, L.*

[Plate II., Figs. 3 and 3 a.]

The colour of the upper wings of this Moth is brown and of the lower wings yellow, with distinct black markings near the edge. The caterpillars feed at night on crops, and are known as "Surface Caterpillars." Soot, lime, or paraffin may be applied as a dressing, or the caterpillars may be picked off with the hand. The caterpillars of two other moths, viz, the Clouded Border Brindle Moth (Xylophasia rurea) (Plate II., Fig. 4), and the Dark Arches Moth (Xylophasia monoglypha), (Plate II., Figs. 5 and 5a), both of which are common in Ireland, commit similar depredations.

Corn Wolf Moth.—*Scardia granella, L.*

[Plate II., Fig. 6.]

The general colour of the front wings of this small Moth is light brown, the hind wings are whitish. It frequents granaries from April till August, laying its eggs on the grains, the inside of which the caterpillars devour. They unite the grains by a silky web. Extremes of temperature affect the caterpillars. If a light be kept burning in the granary the moths will be attracted to it and destroyed. The natural enemies of these insects are Spiders and Bats. The granaries should be kept clean and in repair.

Corn Saw Fly.—*Cephus pygmæus, L.*

[Plate I., Fig. 11.]

This Fly cuts the upper part of the stem of the corn in early summer, and lays its egg in the cut. When hatched the grub eats its way to the foot of the stem, and then cuts through the stem, causing it to fall. It winters as a pupa in the stump, and next summer the Fly appears. The pupæ may be destroyed after harvest by scarifying the land, and burning the stubble, or ploughing it in.

The Ghost Swift Moth (Hepialus humuli), (Plate II., Figs. 7, 7a, 7b), and the Common Swift Moth (Hepialus lupulinus), (Plate II., Fig 8), also attack corn crops, but their ravages are not often serious. The male of the first named insect has a brownish body, with white wings. The upper wings of the female are yellowish, and the lower wings somewhat darker, with a reddish tinge. The Common Swift Moth is of a brownish colour. The caterpillars of both moths feed on roots, and live underground, and the eggs are laid in summer. The caterpillars live two years before becoming chrysalids.

Other insects ravage these crops, such as a Fly (Hylemia coarctata, Fall), the grub of which lives and feeds within the stem of corn. It changes in autumn to a pupa in the lower part of the stem. Another enemy is a Fly (Oscinis vastator, Curt), the maggot of which eats into the stalk. Neither of these insects, however, is recorded as occurring in Ireland.

The caterpillar of the Common Dart Moth (Agrotis segetum, Schiff), also devours corn crops. It will be found described under Cabbage and Turnip Insects, page 8.

INSECTS INJURIOUS TO CLOVER.

Weevil Beetles.—Apion apricans, Herbst; and Apion assimile,
Kirby.

[PLATE L, FIG. 12.]

The Purple Clover Weevil Beetle (Apion apricans), is shining bluish-black in colour. Apion assimile is similar in colour, but smaller. These Beetles cause much injury. They devour the leaves, and the grubs eat the flowers. Crops attacked should be cut early and used as fodder, so as not to run to seed. All refuse should be burnt. There is another species of this Beetle (Apion trifolii), whose habits are similar, but which has not been reported as appearing in Ireland.

Another enemy, which is not, however, properly speaking, an insect, is a Mite (Volvella coronella), the young of which live between the closely arranged leaves, and suck the sap.

The Weevil Beetles (Sitones lineatus and Sitones crinitus), [see page 10, and Plate V., Figs. 1 and 2], also attack clover.

Insects injurious to Cabbage and Turnip.

Plant Louse (Cabbage)—Aphis brassica, L.

The general form of the Plant Lice is well known. This insect is pale green, dusted with white. It infests the under side of the cabbage leaves, from which it sucks the sap. Syringing with water and soft soap, or tobacco water, is found beneficial, or the parts of the plant attacked may be burnt. The greatest enemies of the Plant Lice are the birds and insects which feed on them. [See Insects injurious to Peas and Beans, p. 10.]

Two groups of Flies which feed on this and other species of Aphis may be mentioned here, viz., the Golden Eyes and the Lace Wings. They lay stalked eggs among the Aphides which may be easily recognized.

Plant Louse (Turnip)—Aphis rapa, Curt.

The colour of this insect is brownish. Its habits are the same as those of the foregoing, and the remedies and means of prevention similar.

B 2

Snowy Fly—Aleyrodes proletella, L.

[PLATE III., FIG. 1.]

This minute insect has a yellow body and white wings. It lives underneath the leaves. The Fly sucks the sap, and its ravages may be recognized by whitish patches on the leaves. The best remedy is to cut off the leaves and burn them.

Flower Beetle—Melegethes æneus, Fab.

[PLATE III., FIG. 2.]

The colour of this Beetle is shiny bluish black. It feeds on the pollen, and thus prevents fertilization. The insects may be destroyed by placing bags under the plants and shaking them into them. Early cultivation and dressing the land with soot or lime is found to be serviceable in preventing attacks.

Seed Weevil Beetle—Ceutorhynchus assimilis, Payk.

[PLATE III., FIG. 3.]

The colour of this Beetle is black, with grey tinge. It lays its eggs in the seed pods, and the seeds are eaten by the grub when it is hatched. The Beetles may be caught and destroyed in the manner described above.

Root Gall Weevil Beetle—Ceutorhynchus sulcicollis, Gyll.

[PLATE III., FIG. 4.]

This Beetle is shiny black in colour. It lays its eggs in the root or stem underground. The insect causes galls in which the grubs live, and which sometimes seriously injure the plants. When full grown, the grubs leave the galls and complete their transformation in the soil. The stocks of plants attacked should be dug up and burnt. The application of lime, soot, or ashes to the ground before planting has been found useful.

The Mustard Beetle—Phædon betulæ, L.

[PLATE III. FIG. 5.]

This Beetle is dark blue, dark violet, or dark green. It attacks the leaves immediately they appear. Manuring and early sowing are beneficial, and dressing the ground when wet with lime or soot is advantageous. Another nearly related species is the Phædon armoraceæ L., whose habits are similar.

The Flea Beetle—Phyllotreta nemorum, L.

All weeds belonging to the order Cruciferæ, such as Shepherd's Purse, Charlock, &c. (see Chapter on Weeds), should be taken away and burnt.

The crop should be pushed on when young by good cultivation and manuring.

Another species of this Beetle (Phyllotreta concinna, Marsh), which is of a greenish black colour, also attacks these crops in a similar manner.

Root Fly—Anthomyia radicum, Curt.

[PLATE III., FIG. 7.]

The colour of this Fly is greyish. The roots of cabbages and turnips are attacked by the grubs, which, when full grown, go into the earth and then turn into pupæ, and finally into Flies. Broods appear in succession during summer and till November. The leaves of the plants become yellow, and droop. Lime water sprinkled on the ground is said to be beneficial, and a change to another kind of crop is advantageous. The plants infested by the Flies should be burnt.

Another Root Fly, whose habits are similar, is Anthomyia tuberosa, Curt.

Black Leaf Miner Fly—Phytomyza nigricornis, Mac.

[PLATE III., FIG. 8.]

These Flies are slate coloured. The maggots burrow the soft part of leaves of turnips, just above the under side. The plants attacked should be fed off by sheep, or when stored, the tops which are cut off burnt.

There is no Irish record of this species, but several species of the same genus are Irish.

Yellow Leaf Miner Fly—Drosophila flava, Fall.

[PLATE III., FIG. 9.]

These Flies are of a yellowish colour, with black markings on the abdomen. The maggots mine into the pulp of the leaves just beneath the upper surface, producing blisters and discolouration. Lime, soot, or guano dusted on the leaves when wet is found useful, and after an attack the tops cut off when storing should be burnt.

. There is no Irish record of this species, but several species of the same genus are Irish.

Large White Butterfly—Pieris brassicæ, L.

[PLATE IV., FIGS. 1 and 1a.]

The caterpillars of these Butterflies attack cabbage and turnip. The eggs are laid in clusters underneath the leaves, and may be destroyed. In winter the chrysalids may be found behind old boards, &c. Two broods appear in the year—one in spring and the other in midsummer.

The White Butterfly has formidable enemies in the Ichneumon Flies, which lay their eggs within the body of the caterpillar, and the grubs when hatched devour it internally. They then come out of the body and spin yellow cocoons, from which the Flies afterwards emerge. These cocoons have been erroneously thought by ignorant persons to be "caterpillars' eggs," whereas in reality they contain the caterpillar's most deadly enemies and the farmers' friends.

Small White Butterfly—Pieris rapæ, L.

[Plate IV., Fig. 2, 2a, and 2b.]

This insect is also well known. The colour of the body is black and the wings white, with black spots, the male having less black marking than the female. It lays the eggs singly on the under surface of the leaves of plants belonging to the Cruciferous family. The ravages of this Butterfly are similar to those of the Large White. There are two broods in the year—one appearing in spring and the other in midsummer.

Green Veined White Butterfly—Pieris napi, L.

[Plate IV., Fig. 3 and 3a.]

The wings of this creature are white, tipped with black, and the back black. The male has one black spot, and the female two on the fore wings. The ribs on the under side of its wings are greenish. The egg, which is a pale green, is laid singly on the under side of the leaf. This Butterfly attacks crops in the same way as the two insects last described. There are two broods in the year—one appearing in spring and the other in midsummer. It has some natural enemies in the Ichneumon Flies.

Cabbage Moth—Mamestra brassicæ, L.

[Plate IV., Fig. 4 and 4a.]

These Moths are of a rich brown colour. The caterpillars destroy cabbages, the leaves of which they eat and gnaw down to the heart. The eggs are laid on the leaves and hatched in a few days. The chrysalis remains in the ground in the winter, and the Moths appear in May.

The caterpillars and the chrysalids may be searched for and destroyed. When plants are attacked, spent gas lime is found beneficial.

Common Dart Moth—Agrotis segetum, Schiff.

[Plate IV., Fig. 5 and 5a.]

The colour of these Moths varies from grey to brown, the upper wings being much darker than the lower. When first hatched, the caterpillar gnaws the leaves and stem, and then devours the roots. Before sowing, the ground should be cleared of all weeds, especially charlock. The insects should be searched for and destroyed. Rooks, starlings, ravens, jackdaws, magpies, &c., devour them, and should be encouraged.

Heart and Dart Moth—Agrotis exclamationis, I.

[PLATE IV., FIG. 6.]

This Moth is of a somewhat similar colour to the last-named insect, but has distinct dark markings on the upper wings, by which it may readily be recognized. The caterpillars feed at night, hiding during the day beneath clods and loose rubbish. They gnaw off the tops of the leaves and go from plant to plant.

Clods and weeds which shelter them should be removed, and at night the insects may be hand-picked. Watering the ground with gas water, or digging soot into it, is found efficacious.

Garden Pebble Moth—Pionea forficalis, L.

[PLATE IV., FIG. 7.]

The colour of the upper wings of this Moth is lightish brown, and of the lower wings whitish. The caterpillars eat the leaves into holes. They may be picked off with the hand, and when the plants are wet, sprinkling with fine ashes or dry lime is beneficial.

Diamond-back Turnip Moth—Plutella cruciferarum, Zell.

[PLATE IV., FIG. 8.]

This Moth has upper wings of a brownish colour, varying somewhat in shade ; the lower wings are grey. The caterpillar attacks the leaves of turnips and cabbages, principally the former. It has occasioned much injury in this country. An effectual remedy is to shake the caterpillars from the plants, and then throw lime upon them. Rooks, thrushes, &c., devour the caterpillars ; which have also another enemy in an Ichneumon Fly.

Turnip Saw Fly—Athalia spinarum, Fab.

[PLATE III., FIGS. 10 and 10a.]

The colour of this Fly is black on the head and front part of its body, and light orange behind. Its grub is most destructive to the leaves of turnip plants, sometimes destroying entire crops. During summer a succession of broods appear. The grubs may be brushed off with light branches of trees or furze, the ground being then covered with soot or lime.

Thickly sown and well manured crops are less likely to suffer than weak ones.

Insects injurious to Peas, Beans, Beet, Mangold, and Asparagus.

INSECTS INJURIOUS TO PEAS AND BEANS.

Plant Louse—Siphonophora pisi (Pea Aphis).

These Aphides, which are of a green colour, are known as "Green Flies" or "Dolphins." They suck the sap from the leaves of peas. Dry lime dusted on the leaves when wet is useful. Starlings and other birds devour these insects.

Both this and the Bean Aphis are preyed upon by the grubs of the Hovering Flies (Syrphus ribesii and Syrphus pyrastri). (Plate V., Figs. 15 and 16.)

The Beetles known as "Lady Birds" are also allies of the farmer. The appearance of these insects is well known. Their grubs have six feet, and they feed on the Aphides.

Another enemy of the Plant Lice is an Ichneumon Fly (Aphidius). The female lays her eggs in the body of the Aphis, which is devoured internally by the grub. . The grub when it leaves the skin becomes a pupa, which produces a Fly.

Plant Louse—Aphis rumicis, L.—(Bean Aphis).

These insects are known as "Colliers" or "Blackflies" from their colour, causing the plants to appear as if coated with soot. They multiply very quickly, and suck the sap from the leaves, attacking the top of the plant first. These tops should be at once removed and burnt. Applications of strong soap-suds are beneficial. The Aphides feed also on weeds, such as the curled dock and thistles, which should be cleared away.

Pea Weevil Beetles—Sitones lineatus, L., and Sitones crinitus, Oliv.

[PLATE V., FIGS. 1 and 2.]

The first of these insects (Sitones lineatus), is of a light clay colour with striped back. Sitones crinitus is smaller in size and more of a grey or rosy colour. They gnaw the edges of the leaves of peas and beans, and their grubs are believed to attack the roots. Their ravages are in April and May. The Beetles shelter at night under clods and rubbish. Dressing the ground with lime or soot is found to be effective, and manuring to promote rapid growth beneficial.

These insects are devoured by Starlings. Another species of Pea Weevil Beetle, whose habits are similar, is Sitones puncticollis, Steph. There is, however, no Irish record of its appearance.

Pea Beetle—Bruchus pisorum, L.

[PLATE V., FIG. 3.]

The general colour of this insect is brown. It lays its eggs in the pods of the pea, and the grub injures the seed. The injured seeds cannot produce healthy plants, and they may be known by the dull round marks round them. They should be carefully picked out, as the grubs within will become Beetles which will attack the next year's crop.

Bean Beetle—Bruchus rufimanus, Bohem.

[PLATE V., FIG. 4.]

The colour of this Beetle is dark brown, and its habits are similar to those of the last named Beetle, except that it attacks the bean. It should be treated in the same way.

Both species are liable to attack from Ichneumon Flies, which to some extent keep their numbers down

Pea Moth—Endopisa nigricana, St.

[PLATE V., Fig. 5.]

The colour of this Moth is brown. The caterpillars eat the peas in the pod and then burrow into the ground. Infested plants should be burnt after harvest, and the ground deeply dug. This insect is believed to be rare in Ireland.

INSECTS INJURIOUS TO BEET AND MANGOLD.

Night-feeding Ground Beetle—Pterostichus madidus, Fab.

[PLATE V., Fig. 6.]

This Beetle is of a black colour. Though usually carnivorous, it is said to attack the roots of mangold. Sand, ashes or sawdust, mixed with paraffin, should be applied to the plants as a dressing.

Carrion Beetles—Silpha opaca, and Silpha atrata, L.

[PLATE V., Figs. 7 and 8.]

The Silpha opaca is a little blackish Beetle. The Silpha atrata is black.

The grubs of these Beetles are supposed to feed usually on carrion, but have been known to cause immense damage in Ulster to the leaves of Mangold and Beet. The attacks are in May and June. Farm-yard manure for these crops should be put in in autumn instead of in spring, as the Beetle lays its eggs in April in decomposing matter. If manure is applied in spring it should be well rotted or artificial. Soot, sulphur, or lime may be used after rain for dusting crops attacked.

Beet Beetle—Atomaria linearis, Steph.

[PLATE V., Fig. 9.]

This little Beetle varies from rust colour to black. It appears in May and June, attacking the root, leaves and bud, and causes great damage. The plants should be dressed with soot, lime, or liquid manure.

Weevil Beetles—Otiorrhynchus sulcatus, and Otiorrhynchus picipes, Fab.

[PLATE V., Figs. 10 and 11.]

The Black Vine Weevil Beetle (Otiorrhynchus sulcatus) is a dull black colour. The clay-coloured Vine Weevil (Otiorrhynchus picipes) is smaller. The leaves and buds are attacked by these Beetles, the grubs of which feed on the roots. The ravages of the Beetles are committed at night. By day they hide beneath clods, &c. Dressing the ground in the day time with lime or soot will deter the Beetles returning to the plants; and roots which have been attacked by the grubs should be similarly dressed.

Leaf Beetle—*Gastroidea polygoni*, L.

[PLATE V., FIG. 12.]

This Beetle has the head and front part of the body yellow, and wing-covers bright metallic green. It devours the leaves.

The Silver Y Moth—*Plusia gamma*, L.

[PLATE V., FIG. 13.]

This Moth is of a grey colour, and its name is occasioned by a spot shaped liked a Y in the centre of the fore wings, by which it may easily be recognized. It lays its eggs under the leaves which the caterpillars attack. Dressing the infested plants with soot, lime, or salt is advisable. The caterpillars may be brushed or shaken off and destroyed. Insectivorous birds should be encouraged.

Other insects attack these crops. Mangold is attacked by a species of Plant Louse (Aphis) which should be dealt with in the way described for Pea and Bean Aphides. The Turnip Moth (Agrotis segetum, Schiff), described at p. 8, is also destructive to Beet and Mangold. Another foe is a Fly (Anthomyia betæ Curt.), (Plate V., Fig. 14), the maggots of which feed on the pulp of the leaves of Beet. This Fly is not very common, and is not reported as having been found in Ireland.

INSECTS INJURIOUS TO ASPARAGUS.

Asparagus is injured by a Beetle (Crioceris asparagi L.), the grub of which attacks the leaves and devours the young shoots. The length of this Beetle is about a quarter of an inch, and it is blue-black or greenish in colour. There is no record of its appearance in this country.

Insects injurious to Carrots, Celery, Parsnips, Lettuce, Onions, Potatoes, Radish, and Horse Radish.

INSECTS INJURIOUS TO CARROTS.

Plant Lice.—*Aphis carota and Aphis papaveris?*

Carrots are attacked by these Aphides. The remedies are the same as those for the Pea and Bean Aphides. (See pages 9 and 10.)

Carrot Fly.—*Psila rosæ*, Fab.

[PLATE VI., FIG. 1.]

The red colour known as "Rust" on carrots, is produced by the maggots of this Fly, which gnaw the roots. The roots become shrivelled and die, and the leaves change colour. The grubs go into the earth when full grown, where they change to pupæ and ultimately become Flies. The crops should be thinned when the plants are young, and if Flies appear, the ground should be sprinkled with ashes or sand soaked with paraffin—soot has also been found efficacious. All roots attacked should be burnt in the autumn and the ground deeply ploughed and limed.

Common Flat Body Moth.—*Depressaria applana*, F.

[PLATE VI., FIG. 1.]

This Moth is of a dull reddish ochre colour. The caterpillars injure carrot crops, folding and devouring the leaves. They may be shaken down and killed with lime. Another species is the Carrot Blossom Moth (Depressaria nervosa, Haw.), which feeds on the blossoms. This insect, which is reddish brown in colour, is, however, rare in this country. The Purple Carrot Seed Moth (Depressaria depressella, Hüb.), feeds upon the seeds of carrot. It is not recorded as occurring in Ireland.

The natural enemies of the Depressaria are the Small Solitary Wasps (Odyneri) (Plate VI., Fig. 3), the grubs of which devour the caterpillars.

INSECTS INJURIOUS TO CELERY AND PARSNIPS.

The Celery Fly.—*Tephritis onopordinis*, Curt.

[PLATE VI., FIG. 4.]

This Fly is of a tawny colour. The maggots devour the inside of the leaves of celery and parsnip on which they cause brown blisters. The maggots change to pupæ in the ground, and when the crop has been gathered may be destroyed by digging in fresh gas lime. Infested leaves should be destroyed. Two or more broods of this insect appear during the summer.

There is another Fly destructive to celery, viz :—the Stem Fly (Piophila apii, Westw.) (Plate VI., Fig. 5), the maggots of which gnaw the stem into galleries. There is no record of this insect being found in Ireland. The Turnip Moth (Agrotis segetum, Schiff.), is also injurious. (See Cabbage and Turnip Insects, p. 8.)

INSECTS INJURIOUS TO LETTUCE.

Plant Louse.—*Pemphigus lactucæ*, Westw.

This is a small greenish Aphis which attacks the roots of lettuce, causing the leaves to droop. The plants attacked should be at once removed and the roots destroyed. The ground should be drenched with lime water, soap-suds, or liquid manure.

Other insects attack the lettuce, such as the Silver Y Moth (Plusia gamma, L.), described at p. 12, and the Lettuce Fly (Anthomyia lactucæ, Bouché) (Plate VI., fig. 6), which lays its eggs on the flowers, and the maggot of which devours the seed. There is no Irish record of the appearance of this Fly.

INSECTS INJURIOUS TO ONIONS.

The Onion Fly.—Anthomyia ceparum, Bouché.

[PLATE VI., FIG. 7.]

This Fly is of a reddish grey colour, the male being darker than the female. The maggots sometimes entirely destroy the onion bulb within which they feed. They change to pupæ in the ground and then become Flies. During the summer a succession of broods appear. Attacked plants should be dug up and burnt, and the land ploughed and dressed with lime after the crop has been gathered.

The attacks are more severe on badly prepared land than when well manured.

INSECTS INJURIOUS TO POTATOES.

Frog Fly—Eupteryx solani, Curt.

[PLATE VII., FIG. 1.]

This small Fly is of a bright green colour. It sucks the sap from the stems and leaves, but the injury caused is not often serious. It may be destroyed in the same way as the Plant Lice.

Colorado Beetle—Doryphora decemlineata, Say.

[PLATE VII., FIG. 2.]

The colour of this destructive insect is buff yellow, with five distinct dark lines on each wing-case, by which it is easily recognised.

Happily it has not gained a footing in this country, but as a precautionary measure, it is desirable it should be mentioned here. In 1877 it was found in a steamer at Liverpool. The occurrence of this Beetle should be at once reported to the Police.

The Hawk Moth—Death's Head—Acherontia atropos, L.

[PLATE VII., FIG. 3 and 3a.]

This insect is of a rich brown colour, with the lower wings yellow. It is the largest of British Moths. Its caterpillars eat potato leaves. They are, however, not common enough to cause much injury. They feed in the evening, when handpicking should be resorted to. They pass the chrysalis stage under ground. One brood appears in spring and another in autumn.

Other insects attack the potato crop, such as the Millipede described at page 1, and the Click Beetles at page 2.

INSECTS INJURIOUS TO RADISH AND HORSE RADISH.

The Root Fly (Anthomyia radicum, Curt), described at page 7, attacks the radish, and the Garden Pebble Moth (Pionea

PART II.

MICROSCOPIC FUNGI INJURIOUS TO FARM CROPS.

It is now proposed to consider the second branch of the subject, the more common Fungi which cause injury to our crops ; and the remarks have been divided under four heads, viz. :—those Fungi attacking—1, Cereals; 2, Cabbage and Turnip; 3, Peas and Beans ; and 4, Other Crops.

Fungi injurious to Cereals.

Corn Mildew—Puccinia graminis, Pers.

This Fungus attacks in the summer the leaves and culms of corn and grass, on which it appears in elongated reddish spots ; later on this spot becomes nearly black. It is very common, and frequently causes serious loss. The seed should not be taken from plants attacked, and a change to green crops has been found beneficial. A top dressing of salt has also been found useful. All mildewed straw and grass in the vicinity should be carefully burnt as soon as noticed. Barberries should not be grown in the hedges in the vicinity of corn crops, as a second form of the Fungus occurs on the barberry.

Round Corn Rust—Puccinia rubigo-vera, Pers.

This Rust attacks cereals in the spring. The spots are yellow generally on the upper surface, and the spores are a yellowish powder. The remedies are the same as for the preceding. It is not nearly so injurious to the crop as Puccinia graminis.

Corn Smut—Ustilago carbo, Tul.

The Corn Smut is a familiar Fungus, which in autumn appears on the ears of wheat, barley, and especially oats. It is found also on various grasses. Its spores are black and easily recognized. All cereals attacked and grasses affected by the Fungus should be destroyed. The only effectual means of preventing the attacks appears to be the dressing of the seed, for which there are various receipts.

Bunt—Tilletia caries, Tul.

This Fungus appears on wheat in the autumn and is very common. It fills the grains with dark spores which, when crushed, emit a foul smell. It is difficult to detect at first without close examination. The plants attacked are usually greener than the healthy ones. The diseased grains are generally of different colour, being somewhat greenish ; and they are ordinarily different in shape, being wider and shorter. When the grains crack the dark powder of the spores is seen, which is distinguishable by its fetid odour. ....

This Fungus appears to be sown with the grain, and as in the case of Corn Smut, the only sure means of prevention is the dressing of the seed with the view to the destruction of the spores. Various chemical preparations are recommended for the purpose.

Ergot of Rye—*Claviceps purpurea, Tul.*

This is a well-known parasitic Fungus which appears on cereals, especially Rye, and various grasses, and is very poisonous in its effects. It has a powerful effect on pregnant animals; bread made from ergotised grain is liable to give rise to gangrene. When sifting, the ergots being generally black, can without difficulty be picked from among the grains by the hand; or, as they are usually larger be caught in the sieve.

Ergot is favoured by moisture, and where it appears special attention to drainage is necessary. The flowers of grasses harbour the spores, and where the disease attacks the grass the top should be mown off and burnt. Grasses in the hedgerows should be carefully destroyed.

Fungi Injurious to Cabbage and Turnip.

Cabbage Mould—*Peronospora parasitica, Pers.*

This Fungus appears in summer and autumn as a white bloom on the under sides of the leaves of Cabbages and Turnips. It is common, and grows also on Shepherd's Purse and other cruciferous plants which act as nurses for it.

All affected plants should be destroyed and weeds and refuse burnt. See also note below.

Crucifer White Rust—*Cystopus candidus, Lev.*

This rust attacks in summer plants belonging to the Cruciferæ family, as Cabbage, Radish, Mustard, Shepherd's Purse, &c. It is white in colour, and the stems and leaves are swollen, blotched, and streaked. A change of crops, and the burning of all cruciferous weeds, appear to be the best remedies.*

Club-Root—*Plasmodiophora brassicæ, Wor.*

The disease known as Club-Root in Turnips and Cabbage was formerly believed to be owing to the ravages of insects. Recently, however, it appears to have been satisfactorily established that its cause is a Fungus which attacks the roots. The Fungus is believed to be confined to plants of the cruciferous family, weeds belonging to which should be carefully destroyed. Any plants affected should be burnt, and an alteration of the crops for some time is desirable.

Fungi injurious to Peas and Beans.

Pea Mould.—Peronospora viciæ, Berk.

This Fungus frequently attacks Vetches and early garden Peas, appearing on the under surface of the leaves in brown patches. It is stated that it has been also found on the following plants :— The Bush Vetch (Vicia sepium, L.), the Common Vetch (Vicia sativa, L.), the Slender Vetch (Vicia tetrasperma, Manch.), and other plants.

All diseased plants of Vetches and Peas should be burnt. See also note, p. 16.

Pea Mildew.—Erysiphe martii, Link.

This Fungus appears on both sides of the leaves of Peas, Beans, Umbelliferæ, and other plants in the autumn. It is at first white in colour, but later on minute yellow and black spots appear. It is stated it has been found on plants belonging to four different families. The only remedy appears to be the destruction by fire of the plants affected, and the clearance of weeds which would nurse the spores.

Bean Rust.—Uromyces fabæ, Pers.

This Rust attacks Beans in August and September, covering the leaves and stems with brownish rust. The remedy is the same as in the foregoing.

Fungi injurious to other Crops.

Lettuce Mould—Peronospora gangliformis, Berk.

This Mould appears on Lettuces and other plants of the Composite family. Pale patches of decay are seen on the leaves on which spots of the Mould may be distinguished with the naked eye. The Fungus grows inwards, attacking the outside leaves first. Frost checks the growth of this Fungus, and when it appears on Lettuces in a frame in early spring the exposure of the plants to the frost for a short time is beneficial.

All weeds of the Composite order are shelters for the disease, and they, as well as decaying refuse, should be burnt. See also note, p. 16.

Onion Mould—Peronospora Schleideniana, DeBy.

The Mould, which frequently appears on Onions, is occasioned by this Fungus. It chiefly attacks the leaves which are covered with a grey-white bloom. Deep trenching is recommended ; and sowing the seeds in autumn, so that the plants may have attained sufficient strength to resist the Fungus when it appears in the spring. The diseased plants should be destroyed. See also note, p. 16.

Parsnip Mould—Peronospora nivea, Ung.

This Mould attacks Umbelliferous Plants ; but its chief mark is the Garden Parsnip. The leaves of the Parsnip are first attacked, and the roots then become diseased and putrid,

All plants attacked should be burned, and all weeds belonging
to the Umbelliferous Family in the vicinity destroyed. Parsnips
should not be grown again in the same place for some time. See
also note, p. 16.

Spinach Mould—*Peronospora effusa, Grev.*

This Fungus attacks Spinach and Goosefoot, and some other
plants. It is of a light purple-grey colour, and appears in spots
on the under surface of the leaves. The remedies suggested are
the destruction of the plants diseased and of any Goosefoot near.
See also note, p. 16.

Beet-leaf Rust—*Uromyces betæ, Pers.*

Yellow spots appear sometimes on the upper surface of Beet
leaves in August and September. These are caused by the
above Fungus. Dr. Cooke reports having found it on the leaves
of the Wild Beet (Beta maritima). The destruction of the plants
affected is the best remedy.

Potato Mould—*Peronospora infestans, Mont.*
Phytophthora infestans, De Bary.

The terrible Potato disease, with which this country was visited
in 1845, is caused by this Fungus, the nature of which has recently
been fully discussed. New varieties of Potatoes have been intro-
duced with the view of securing one less liable to the attacks of
this pest. See also note, p. 16.

PART III.
WEEDS.

LIST OF THE PRINCIPAL PLANTS USUALLY CONSIDERED AS WEEDS.

1

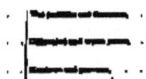

Poisonous Weeds.

Many weeds are prejudicial if eaten by farm animals on account of their acrid taste, but there are several, which are named in the following list, the presence of which must be regarded with suspicion owing to their poisonous properties. Though some of these weeds are not very common, they have been described as a precautionary measure.

Corn Cockle—*Lychnis githago.*

The seeds of this plant are poisonous. The death of six cows fed upon meal largely composed of the seeds of Corn Cockle has been recently reported from Belgium. [For description see List, Page 20.]

Cowbane—*Cicuta virosa.*

This plant is found in wet ditches and near water. The stem attains three or four feet in height. It bears white flowers in summer.

Hemlock Water Dropwort—*Œnanthe crocata.*

This plant is a perennial found in similar situations to the foregoing. It is stout and branched, and attains from three to five feet. The juice becomes yellow when exposed to the air. It flowers in summer.

Fool's Parsley—*Æthusa cynapium.*

[For description see List, Page 21.]

Common Hemlock—*Conium maculatum.*

An annual or biennial plant, usually three to five feet high, emitting an unpleasant smell when bruised. The stems are spotted. It is usually found in waste places, ruins, and along hedges and the borders of fields. It bears small white flowers in summer.

Common Henbane—*Hyoscyamus niger.*

A coarse erect annual, attaining about two feet high, and having an unpleasant odour. It is found in waste and stony places and hedges. It bears pale yellow flowers in summer.

Bittersweet—*Solanum dulcamara.*

This plant is a perennial, and usually found along hedges, river sides, and in other moist places. It attains many feet in length, and bears small blue flowers in summer, and red berries.

Deadly Nightshade—*Atropa belladonna.*

This is an erect perennial, found in waste and stony places. It bears large purple blue flowers in summer, and large shining black berries. It is rare in Ireland.

D

Purple Foxglove—Digitalis purpurea.

[For description see List, page 24.]

Cuckoo Pint, or Lords and Ladies—Arum maculatum.

This plant grows under hedges and in woods, &c. It flowers in spring, and bears in August bright red berries on a spike, partly hid by a sheath.

Meadow Saffron—Colchicum autumnale.

This bulbous herb grows in damp meadows and pastures. It is rare in Ireland. The flowers, which are reddish purple or sometimes white, appear before the leaves in autumn.

Common Darnel—Lolium temulentum.

This plant is found in fields and waste places. It is an annual, and flowers in summer. Its seeds are poisonous.

Though not a native of Ireland, it may be desirable to mention also the Monkshood or Wolfsbane (Aconitum napellus), as several exotic species are commonly cultivated in our gardens. It is an erect perennial, growing to about two feet high, and bears a large blue flower in summer. The root in which the poison is chiefly contained has been taken in mistake for horse-radish, and led to several cases of accidental poisoning.

Extirpation of Weeds.

The consideration of this subject may be divided into two heads :—

    1. Destruction ;
    2. Prevention.

Destruction.—Under this head we have the practical problem of how to get rid of the weeds already in the soil whether as plants or seeds.

The plants are of two classes—Perennial weeds and annual weeds.

Perennial weeds being rooted in the soil present the greatest difficulties. Ploughing, harrowing and scuffling clear the land partly, but these operations do not reach the roots of Coltsfoot Bindweed, Couch, &c. For these the three-pronged fork is the best implement. Docks and Thistles should be pulled up by the roots. Some weeds are usually found in patches, such as Stinging Nettles, in places where they cannot readily be dug up. In such cases the best plan is to mow them down regularly when the leaves begin to show, which will ultimately kill the roots. Alkali waste is useful as a dressing to clear the ground of deep-rooted weeds.

As regards annual weeds, the seed of which is in the soil, the great point is to bring the seeds to the surface and cause vegetation, so that they may be cleared away before the crop is put in. With this object the ground should be ploughed, harrowed, and rolled, and when the weeds appear, ploughed again and again as each instalment of weeds comes up.

In turnip fields weeding should be done with the horse hoe or common hoe, and no weeds allowed to seed.

Prevention.—Weeds are sown in three principal ways—(1) with the seed, (2) with manure, and (3) by the wind—and the measures for prevention must be directed to meet these three cases.

It is a notorious fact that bad seed is a fruitful source of the appearance of weeds; agriculturists should, therefore, be careful to obtain their seeds from dealers of known character.

Weeds are frequently sown, also, with the manure, having been previously thrown on the top of the dung heap, and left there to vegetate. To prevent this, they should be burnt or put at the bottom of the heap to rot.

The seeds of weeds, again, are wafted with the wind; the farmer should, therefore, be careful to remove the weeds from all waste places, hedges, and ditches, as the seeds from these weeds will sow themselves over his fields.

Of late years agricultural implements have been much improved, and this has facilitated the work of the farmer in the eradication of weeds.

Visitors to Ireland have been much struck by the neglected condition of many of our farms in respect of the unchecked growth of weeds, and it is earnestly hoped that, as the serious damage they inflict becomes better understood, energetic measures will more generally be taken to remove this reproach from our country.

DUBLIN CASTLE,
10th October, 1890.

SIR,

I have to acknowledge the receipt of your letter of the 8th instant, forwarding, for submission to His Excellency the Lord Lieutenant, the Report on Insects, Fungi, and Weeds Injurious [to Farm Crops, referred to in page 11 of the Agricultural Statistics of Ireland for the year 1889.

I am, Sir,
Your obedient servant,
W. RIDGEWAY.

The Registrar-General,
    Charlemont House,
        Rutland Square.

DUBLIN: Printed for Her Majesty's Stationery Office,
By ALEX. THOM & Co., (LIMITED), 87, 88, & 89, Abbey-street,
The Queen's Printing Office.

www.ingramcontent.com/pod-product-compliance
Lightning Source LLC
Chambersburg PA
CBHW022026190326
41519CB00010B/1615